Taylor, Keith, 1952-

Life science : and other
stories /

Life Science

AND OTHER STORIES

AND OTHER STORIES

BY

KEITH TAYLOR

HANGING LOOSE PRESS
Brooklyn, New York

Published by Hanging Loose Press, 231 Wyckoff Street, Brooklyn, New York 11217.

Printed in the United States of America

10 9 8 7 6 5 4 3 2 1

Acknowledgments: Some of this work first appeared in the following publications: *The Alternative Press, Arms Akimbo, Artlight, The Bridge, Caliban, Hanging Loose, The MacGuffin, The Michigan Quarterly Review, Notus, Witness* and *The Wooster Review.*

The author wishes to thank the National Endowment for the Arts for a Fellowship in Creative Writing that provided the time necessary to complete much of this book. Hanging Loose Press thanks the Literature Programs of the New York State Council on the Arts and the National Endowment for the Arts for grants in support of the book's publication.

Cover art by Ann Mikolowski. Design by Caroline Drabik.

Library of Congress Cataloging-in-Publication Data

Taylor, Keith
Life science / Keith Taylor.
p. cm.
ISBN 1-88241-314-8 (paper). —
ISBN 1-88241-315-6 (cloth)
I. Title.
PS3570.A9418L54 1995
811'.54--dc20 95-9872
CIP

Produced at The Print Center, Inc., 225 Varick St., New York, NY 10014, a non-profit facility for literary and arts-related publications. (212) 206-8465

Contents

I Through a Crack in the World

II Something Austere, Gentle

III Life Science

IV The Blood Trail

for Christine
…something new…

I
Through a Crack in the World

The Black Crayon

There was just one time. When something happened. He was in the basement, drawing. He was eight and still remembers it clearly. Completely. He remembers it all. His parents would sit upstairs after dinner and watch tv. He would go to the basement and draw. For hours. Until his mother called and he had to go to bed. He loved it down there, with the furnace going on and off. The bare light bulb.

On that night he was drawing a farm. He'd used a black crayon to outline all of the buildings and animals. He enjoyed giving everything a stained-glass look. Then he started at the bottom, coloring the ground and the animals. Chickens. Pigs. Then cows. He colored the sky, and it was always blue. And cloudless. He never drew those puffy white clouds. Never. His skies were blue, unless it was night, of course, and then they were filled with gold stars. He saved his gold crayon for the stars.

After he finished the animals and the ground and the sky, he went back to the buildings. They were all red. He was careful to keep the red inside the outlines. If it got away and started smearing, it would ruin everything. When he finished the buildings, he looked for the black crayon to do the roofs. His crayons were all thrown together in a shoe box, but he always seemed able to find the right one. But this night he couldn't find his black crayon, even though he'd used it not long before to draw the outlines. He tested each one on an old piece of newspaper. The black one wasn't there. He picked up the paper he was drawing on and looked underneath. He got down on his knees and looked under the card table. He checked under the pool table, his father's workbench, around the furnace, and way off in the back corner by the washer and dryer in case the black crayon had started rolling. He couldn't find it anywhere.

He came back to his table and looked down on all his roofless buildings. He thought of brown or green roofs, but knew they wouldn't work quite the same way. He wasn't upset, really. He just looked off into the dark corners of the basement, and when he looked back it was there. The black crayon. Right in the middle of the paper he was drawing on, as if it had always been there.

Beautiful Cause

I am at a conference—"Ezra Pound and the Modernist Epoch"—in a provincial Italian town. The conference claims to be an international retrospective, and the major speakers are all well-known Poundians. This conference, however, wants to be different; before and after each lecture unknown people from obscure places have been asked to give ten-minute talks on some aspect of Pound's life or work. I have been asked to speak after the lecture by a famous critic whose work I admire. I am flattered and very nervous.

For some reason I have determined that there will not be a frank discussion of Pound's politics at the conference. I have nothing new to contribute to the argument but have decided that I will risk the usual platitudes. Finally, I plan to say, we must forgive Pound his fascism, his insanities, his cruelties, because his marriage of music and sense becomes a truth greater than any political statement.

I have slept poorly for several weeks, rehearsing in the dark the quiet control I hope to establish over the emotional pitch of my talk.

The critic does not look like I expected her to look. I thought she would be plump, smiling, energetic and fashionably dressed. Instead, she is lean, her gray hair severely cut, and she never smiles during her lecture. Nervous about my talk, I do not concentrate on the development of her argument. I am only reminded of her lecture when I realize that a kind of tension is rising in the room, that feet are beginning to shuffle, that there is a quiet murmur without words, that the critic seems flushed. She is talking about Pound's politics but is not trying to explain them away as an aberration of genius. She is justifying them, saying that

Pound reacted correctly to the problems of his time, saying that his solutions not only were but *are* the correct solutions.

I expect the audience to rise up, to shout her down. But she is moving quickly, jumping from side to side, darting away from the podium and then returning for a quick look at her notes. Pound is almost forgotten. And suddenly I realize that most of the audience agree with her. She begins to refer to Pound's *bella causa* and soon the audience is murmuring the phrase. *Bella causa; bella causa.* Soon there is a chant. Louder, then louder. Then people are standing—not all, but most—and flashing the stiff-armed salute, jabbing their open palms forward to emphasize the first syllable of each word. *BEL-la CAU-sa; BEL-la CAU-sa.* And the critic has finished. I am next.

Gypsy Fragments

1.

In the center courtyard of the Louvre, American visitors caught two barefoot gypsy girls. The girls were trying to pick the American pockets. A Frenchman, with *Le Monde* rolled under his arm, ran over to help. He slapped at the girls with his newspaper, demanded they return everything, then slapped them some more. The girls howled. Tears streamed down. The man gave them one last swat, and they ran off toward the river.

An hour later, I saw the man with the newspaper and the gypsy girls on a bench in front of the French Academy, speaking Romany and laughing.

2.

About twenty of us—mostly French, a few Germans, a couple of Canadians—had gathered in a cave above the beach outside Sête. Two gypsy men joined us. We ate soup, raw shellfish, and passed bottles of cheap wine. The gypsies sang. Slow songs, and they cried.

"It's Spanish," someone said, "but I don't understand the words."

When they finished, we passed them the last bottle. They were still crying, and they drank all the wine.

3.

My friend and I walked north out of Tuam looking for the place that advertised horse-drawn tinker's wagons for rent. We turned in when we saw a large purple one behind a rundown cottage. In the back a group of tinkers was gathered around a fire.

"Sorry," I said. "I thought this was the place that rented wagons."

They all sat there, looking right at us, glumly.

When we turned away, a woman said, "I'll sell you this one cheap, but not the horse."

4.

The Greek women in Argos all wore black and were off the streets by nightfall. The gypsy women swaggered through the town, swinging their arms and sending their bright skirts swirling, both day and night.

Our Greek friend told us that when she was a child her grandmother said, "Be good, or the gypsies will carry you off in the night."

When we stopped to take a picture of the gypsy camp outside town, an old woman with a purple shawl over her head pushed out her arms toward us and spat.

"The evil eye," our friend said. "She's cursing us."

5.

Outside the gypsy church in Saintes Maries de la Mer, a woman with silver earrings and long strands of silver beads, carrying a naked baby who snored, stood before me, begging. "Pour le bébé," she said. "Pour le bébé." I gave her two francs. She smiled and walked away, shaking her head.

Later that week, in the Camargue, a gypsy man took my friend and me out on barely broken white ponies. We'd told him we knew how to ride. When we arrived at a long straight dike, he let his horse gallop. Ours followed. My friend's horse began bucking while at full speed, and she flew off onto the path. My horse stopped abruptly, sending me up and over its head. I almost landed on my friend. The gypsy man came back and looked down. He was laughing at us. I wanted to be angry, but my friend and I were comfortable on our backs, looking up at the gray clouds above the marsh. Soon, we were laughing too.

My Books and the Philosopher

Everyone at the language school knew that if we wanted to see the philosopher we could go a couple of blocks down to the intersection of the Boulevard Edgar Quinet and the Boulevard Raspail, over by the old cemetery. He would often sit for hours in a small café there, holding court or scribbling away in a pocket notebook. But we felt reluctant. We wanted to be as casual about him as the French were, as if there were nothing spectacular about the only man who had turned down the Nobel.

He was still a dominant force in the intellectual life of the city, even though his ideas were already falling from fashion.

So we waited, not mentioning him, until the time every week or two when he and his companion shuffled down the Boulevard Raspail and passed in front of the school. We all watched. The old couple never looked as if they were out for exercise, although none of us could offer any other explanation for what was plainly an unpleasant excursion. The woman, whose books were more influential at that time than were his, looked severely ahead, obviously forcing herself to walk slowly so the philosopher could keep up. He looked down, and we could see only the thick rims of his glasses. He didn't hold her arm, although he looked frail enough to need the support. He clutched his cane tightly.

When my classes were over for the day, I often stayed in that quarter because the woman I was seeing lived close by, up the Rue Delambre from the big intersection. The four cafés were, of course, already famous and far too expensive for us. The only time I went into Le Dôme was when a man dropped dead beside me while I waited for the light to change on the Boulevard Montparnasse. The man gagged, fell back into one of the outdoor chairs and turned blue. I loosened his tie and shirt. A waiter tried to revive him but quickly gave up. I helped carry the body inside behind the bar and waited until the gendarmes arrived so I could give my statement.

I don't think the old philosopher and his friend went into those cafés anymore. At least I never saw them among the serious looking people behind the glass windows.

But it was still their quarter, and everyone knew it.

After I decided to return to America, I had to work out an elaborate way to mail my books. I would pack up small boxes of them, never more than ten titles, back in my room on the Rue du Caire, over on the right bank, in the red light district close to the Porte St. Denis. When my friend came by in her Deux Chevaux, I would take a few boxes to her side of the city. After she found a parking space, I would carry the boxes of books down the Rue Delambre, cross Raspail, and go to a small post office a couple of blocks east on Montparnasse.

As the date of my departure grew closer, I became more and more frantic about my books. I would load as many of the small boxes as I could into my friend's car. Back in her part of town I would carry them through the streets, piled up to my chin. I broke into a good sweat long before I arrived at the post office.

It was on one of the last of these trips, just at the point where I was passing behind Le Dôme and was crossing the sidewalk on the Boulevard Raspail, that I bumped into him. The philosopher came from the left. I didn't see him.

I didn't hit him hard, and I staggered dramatically away from him to avoid any damage. All the boxes fell and three broke open. Luckily the old man was unhurt and still standing, even though I sent his cane flying out into the street.

I knew who he was, immediately. Those fish eyes of his darted around behind the bowl-like lenses. He never really focused on me. I don't think he could.

I dodged cars and recovered his cane. He mumbled

something I couldn't understand, although it sounded like a curse, and I apologized profusely. He shuffled off down the Boulevard Raspail. It was the only time I saw him without his companion.

Bookseller

I don't really care about the characters in books, certainly not about the people who write them. I'm not much interested in their contents, how one book supports or argues with another. I like the objects, the shape and color of them, and the pleasure they take in themselves.

They create abstract patterns: a large red book separated from a small blue one by a used yellow; or a whole shelf assuming twenty shades of green, the variations blending into each other until individual volumes almost disappear. The books make an ever-evolving kaleidoscope, colors shifting into new patterns almost every day.

I prefer to think that it's willed, chosen by the books themselves.

But there is a deeper joy in the bookshop, one that comes despite or because of my need to keep the books in some kind of order. A book on tantric meditation will suddenly appear in the psychology section. I reshelve it. A month later it pops up in the European history section. Six months after it's in the last shelf of the poetry books, looking comfortable between Yeats and Zukofsky. The books find their own order. Their movement seems a dance with a geologic tempo, so slow I can't see it. The dance is like the story country children tell about trees moving at night, just a millimeter or two, nothing that can be noticed in the morning, until one day a child climbs onto a rope swing that's tied to a maple branch, and he swings out as he has done hundreds of times before, but this time slams straight into the side of the barn.

The Customer

She'd been a customer at the shop for a few years, not a major one, but a regular who bought a book or two a month. Mostly the $2.98 or $3.98 remainders we kept on tables outside. Inside, she bought novels, mainline middle-of-the-road fiction, the kind that gets to the bottom of the paperback bestseller list and stays there for a couple of weeks. She always paid in cash, so I never learned her name from checks or credit cards. I was polite when I waited on her. I'm always polite.

She had spectacular hair, black and thick and shiny. It was the kind that couldn't be cut into a fashionable style, unless she cut it very short. She kept it clean and brushed and often let it fall wherever it fell. Sometimes she'd pull it back in a loose ponytail, with strands coming free and blowing around the edges of her face. I figure she might have been Greek or Italian, something Mediterranean.

Early last fall, I started walking home instead of taking the bus. I walk most of the year. But in summer I sweat too much. I like walking in the fall, watching leaves change, then drop. I still enjoy kicking the leaves that collect in gutters or on the edges of sidewalks. In the fall I'll often walk several blocks out of my way, just for exercise.

That first time, after I'd punched out, I saw her a block ahead, walking north. I started in the same direction. She walked four or five blocks, then turned left down one of the small side streets lined with old Victorian houses subdivided into over-priced apartments. By the time I turned after her, she was gone.

The next day I walked the same way but didn't see her. The third day, just as I was going down the small street, I saw her coming out of the fourth house on the left. She locked the door. When she passed me on the sidewalk, we exchanged a hesitant nod.

I didn't walk that way again for over a week.

As the weather grew colder and the first snows began, I started walking down her street regularly. Not every day, but at least two or three times a week. Usually I didn't see her. Once, though, I was right behind her, a hundred yards or so. I figured that if she noticed me, she would assume I was on my regular route. But it was cold. She wore a heavy coat, the fur-lined collar pulled up around her ears. Her hair looked wild and wind-blown, some of it under her coat, some flying out above her collar. She turned quickly into her house. Just as I was passing in front, I saw the upstairs light come on. She was in the room, and I could see her clearly. She took her coat off and threw it on a chair. She shook her hair out and brushed her right hand back through it. I stopped and looked up. Just for a second.

As the days grew shorter, it was dark, almost night, when I walked home. I continued passing her house a couple of times a week. Her light was usually on, and the shades were drawn. Sometimes I found myself on the sidewalk looking up at her window. But I tried not to, tried to keep walking. Casual.

The heavy snows started a few days before Christmas. Most of the students who lived in that neighborhood had already left for vacation. She had been walking home, again, a block or two in front of me. Snow twisted around the street lamps and drifted on the sidewalk. When I passed in front of her house, I looked up. The window was still dark. I stopped. The light came on, suddenly illuminating a new swirl of snow. I caught a glimpse of her. She glanced toward the window, and she looked frightened. Then she quickly reached over and turned out the light. I sensed that she moved closer to the window and thought I heard her say something. Maybe she yelled.

I stood for a moment, then walked away quickly. It was still snowing. I never walked that way again.

In a Garden at Mysore

(for A. K. Ramanujan)

A bird sings in a garden at Mysore. It is late afternoon. The bird's song has five notes, with a trill between the second and third. Occasional variations keep the song from becoming monotonous.

In the back room of a house behind the garden, a young woman and a young man, recently married, lie naked on a rug. The young man's hand rests on the thigh of the young woman. His hand is calloused, and, even though he has washed carefully, there is still grease beneath his fingernails. The young woman's skin is soft and smells like tea.

The young man grew up in this house. His parents gave the back room to the young couple after the marriage. His father is a minor official in the municipal government. The young man, slightly rebellious, has been fascinated with internal combustion engines since he was a child. His father helped him secure a position working on the cars and trucks that belong to the city of Mysore. The young woman, originally from a village a few miles out of the city, stays home and does needlework in the back room. A friend of the family, a merchant, sells her work for a few rupees. The two young people save what money they can. They hope to buy a television or a motorcycle.

Occasionally bored, occasionally angry, they still forgive each other for their boredom and their anger. No one they love has died. Neither of them can speak before the beauty of the other's body.

When the young man dozes, the young woman rises and goes to the window to see the bird that sings in the garden. The bird is hidden in a tree covered with red blossoms. The young woman cannot find the bird, but through the branches of the tree she sees the palace, built by one of the last princes of Mysore and rising above the center of the city, glowing orange, almost golden in the dying light. The young man mumbles and turns but does not awaken. Behind the palace is the outline of the mountain, purple in the evening. Rising over the city, it looks like a giant elephant, asleep.

The Door in the Woods

On state owned recreation land in the western part of the county, the Department of Natural Resources maintains several nature trails. They wind in and out of various habitats: swamps, a bog, a beech-maple forest, oak-hickory forests, old fields. The trail farthest from any of the roads ends at a large wooden barrier that blocks the path and extends for about fifty feet on either side. In the middle is a door, perhaps eight feet high, that looks as if it could have been used in old movies about lonely prairie forts in the middle of Indian country. There are no signs indicating what may be behind the door, and the barrier is not insurmountable. Although it would be difficult, any child or curious adult could figure out a way around it. But there are no irregular paths or trampled weeds showing that people have tried. It seems as if whoever walks this far gets to the door and just turns around.

About thirty years ago and a long way from here, out in a national park on the western edge of the country, a group of hikers discovered a high glacial cirque, a hidden basin that wasn't discussed in any of the guidebooks. It didn't appear on any of the maps. They thought they might be the first people to have found it. The basin had the expected clusters of subalpine firs, the scrubby little trees that grow slowly and almost miraculously up to the edge of timberline on many of the western mountains. A subalpine fir six inches wide may be more than a century old. But those hikers found the largest subalpine fir anyone had ever seen: one hundred and thirty feet tall with a circumference of more than seven feet at the base. When they walked around it, they found on the back a small but solidly attached and perfectly finished door.

II
Something Austere, Gentle

The Footwashing

When Reverend Alvin Traub, the District Superintendent of the province, visited the Didsbury church, he celebrated communion. After the grape juice and biscuits, he officiated at a footwashing.

Reverend Traub explained the theological foundation of the ritual. It was based on an episode in John's Gospel that is not mentioned in the other records of the famous supper. John says the Lord bent down beside the table, poured water into a basin and washed the feet of the disciples, even the feet of Judas just moments before he turned and fled for his silver and the tree. Peter tried to refuse, then said, "Lord, not my feet only, but also my hands and head."

The footwashing created arguments in our house. Mother hated it. She had to go to a room with the other women, wrestle with hose and garter and girdle, then come back to the sanctuary with pale, bare legs. She worried about her foot odor. She didn't want to touch the feet of the old ladies, and she dreaded kneeling in the puddles that inevitably collected on the floor of the church. Besides, she was ticklish.

But Father enjoyed the service. He liked to watch the deacons scurrying around with little granite basins, emptying dirty water and coming back with clean. They were careful not to slop before the altar.

He remembered his initiation to footwashing, the time he first met Alvin Traub. Reverend Traub was a former farmer and carpenter who had educated himself to ordination and the presidency of the Bible College before being appointed the Superintendent of the entire District. He had become a legend in our church. Reverend Traub came down the aisle, touched my father's shoulder, and said, "Son, I would like to wash your feet."

Father was only eighteen, in his first year of ministerial studies, and was very uncertain of his calling. He remembered shaking so hard he had trouble removing his socks and rolling up his pants. The old man knelt before him and gently lifted his foot into the basin.

When Reverend Traub finished, Father offered to return the ritual. The old man's feet were the largest my father had ever seen. They were covered with bunions. But Father's clearest memory of that day was that when he lifted one of those gigantic feet and placed it in the basin, he felt as if he were lowering a newborn animal, a calf, maybe, or a lamb, down into water for its first swim.

WCTU

Lyle Stauffer always won. We memorized long poems about the horrors of drink, about fathers who beat mothers then disappeared for days into smoky taverns, about brave eight-year-olds who wandered streets filled with loose women and drunks until they found Dad and brought him home to the little white house. Father would kneel before mother, cry, and swear to the Highest Authority that it would never happen again.

We practiced our gestures. We learned to break our voices at the moment when the child found father slumped in his barroom chair. She shook the sleeve of his jacket until he looked up and, in his stupor, slapped her face. A drop of blood in the corner of the child's mouth. A faint recognition would come, finally, into father's red eyes. In a blinding moment of repentance he rose and followed the child out into the pure night air.

We recited our poems for the local chapter of the Women's Christian Temperance Union. Some of us forgot the words and stumbled and had to repeat whole stanzas. We knew the winner would be the one who could make Auntie Edna Eby weep.

Lyle never forgot his words. He could make Auntie Edna sob out loud. He always got the gold pin and the chance to move on to the provincial contest in Red Deer where he might get a write-up in the church magazine or a scholarship to the Bible College.

Resurrection

She rose from the grainy black and white photograph in *The Calgary Herald,* a slavic Venus, her blonde hair falling down her back. I had never seen anything like her. I tore the picture out and studied it for hours in the granary. A Doukhobor woman. Completely naked, with more naked Doukhobors blurry behind her.

She had stripped before the Parliament Building to protest government intrusion into the life of the Spirit. I knew Doukhobors did this, but I could never understand their reasons. The few times they came to town, they were dressed plainly in dark homespun. The men wore black hats, and the women had white gauze prayer bonnets just like the Mennonites. And the Mennonites were practically like us. My grandmother spoke respectfully of the Doukhobors, and told me about what she called "their great faith."

The name of their sect meant "Spirit Wrestlers." Seventy years earlier the Russians had derided them with that name. The Doukhobors knew that fighting with angels or with the Spirit was dangerous, that it would probably wound but not necessarily kill. They kept the name even after the pogroms started, when thousands were whipped and hundreds killed, when the Cossacks came and took their children.

Tolstoy, old and striving for sainthood by that time, grew so angry about the Czar's treatment of the Doukhobors that he abandoned all his resolutions and returned to fiction, finishing one last novel. *Resurrection.* It had another of those characters the Count modeled after himself. In the novel Nekhlyudov, a self-tortured member of the aristocracy, is called to jury duty where he must sit in judgment on a prostitute accused of murder. He recognizes her as a former servant he had seduced years earlier, long before he realized that human beings must struggle constantly to be good. The woman is innocent, but through legalistic bungling is sent to Siberia for four years. Nekhlyudov sells his lands and follows the woman into exile.

Tolstoy resolves the whole thing with Bible verses. He was in a hurry. He needed the money. His wife, Sonja, was clinging tightly to the family purse strings. Over the years she had learned too much about the Master's generosity. He needed a new source of income if he were to help the Doukhobors. He finished *Resurrection* during the final years of the century. The Russians went crazy for it.

Tolstoy gave the royalties to the Doukhobors so they could afford to emigrate to Western Canada, where the local governments didn't learn how to handle them for over half a century. When I was born the Doukhobors were actually blowing up bridges and roads because they thought that bridges and roads prepared the path for the Anti-Christ. They calmed down over the next few years, even though they would still show up occasionally to disrobe on the lawns of the various provincial parliaments where the newspapers would snap photographs.

My family were devout members of a rural Mennonite splinter group. Our church had begun to forget its shared theological root with someone who thought she could do the Lord's work by taking her clothes off in front of other people. The Doukhobors were embarrassing. But, even though our family tried to shield us, they couldn't control the newspapers, couldn't save all of us from one perfect body, without blemish in newsprint, burnished by the afternoon sun shining unhindered on a distant government lawn.

What the Angels Said

Matthew Yoder did nothing remarkable until many years after his ordination, when he became the first minister in the history of his denomination to perform exorcisms. There was a great deal of discussion about this at church headquarters in Fort Wayne, Indiana. Matt had placed his leaders in a theological quandary. They believed the Devil played an active role in human affairs, even if they were suspicious of anyone who actually claimed direct contact with Him. But the Devil and the different ways to control Him had a popular appeal that year, so Matt was allowed to continue his exorcisms, although he had to make it absolutely clear that he was exorcising on his own and not as a representative of the denomination.

After exorcising one or two demons a year for several years, Matt began to hear voices. He believed the voices came from visiting angels. He mentioned the angelic voices in his sermons but did not claim that he was speaking with divine authority. Nonetheless, some of the members in his small southern Michigan parish began to suspect that if they were not hearing the Word of God they were at least listening to someone very closely placed. After a few months of this, however, a Sunday morning visitor from Fort Wayne reported to the church authorities that Matt might have moved a bit past what they considered theologically rational.

The church leaders who had allowed the possibility of demons possessing certain members of their ecclesiastical flock couldn't accept the possibility of the Divine communicating directly and in English with one of their preachers. Matt lost his ministerial credentials, was forced to leave his church, and soon disappeared with his wife and children.

They did not leave a forwarding address. Several months later, someone wondered if anyone had ever asked Matt what, exactly, the angels said.

Everything I Know

When I first went to Scotland, I planned to spend two or three weeks sitting beside Loch Ness, looking at the water, waiting to see a mysterious ripple or the tip of an inexplicable fin break the surface. But when I got there, I heard about a tropical garden up along the northwest coast, a placed warmed by the gulf stream smacking against Britain and kept constantly wet by the mist, a miniature Amazonia across the Highlands.

A delivery truck picked me up on the road outside Inverness and took me for two hours along the bay and up into the hills. The road got progressively smaller, but was still paved and fairly smooth. At an unlikely intersection, two roads crossing each other in a high valley filled with heather, the truck turned north. I got out and stood on the empty road leading off to the west. There were no cars, so after half an hour I set off walking through the valley and up into the hills beyond.

After walking for several miles without seeing a car, I was tired. My pack felt heavier. I took it off and sat beside it on the edge of the road. The bare Highlands were all around me. No trees or bushes, no farms, no people.

I was staring back down the little road toward the east, thinking about nothing, just feeling good about being so far away from any place I had ever known. And then there was a little man in front of me, fifteen or twenty yards back down the road. I had been looking there, a bit absentmindedly but still looking. He wasn't there, and then he was.

He was about four feet tall, but in perfect proportion, built more like an athletic child than a midget. He had long red hair and a long red beard. And he was wearing a green hat, the kind working men wear in southern Europe. He and I looked at each other and smiled. I almost waved. And then he was gone. No puff of smoke. Nothing dramatic. He just disappeared, out there in the Highlands, where there were no trees, no ditches, where the air was very clear, and where I could see everything.

Mennonites at Wind River

(for A. L. Becker)

Two cars with Ohio plates pull in at the gas station outside Wind River (The Coldest Spot In Canada—72 Degrees Below Zero!). In the first, a Mennonite man—hair cut close, his beard long and untrimmed, bib overalls, his neck and hands hard and dark. In the second car, his wife—an austere brown dress, heavy-soled shoes, a white prayer bonnet holding in her profusion of hair.

They've come north to collect the body of their son, a pious young man and a help at harvest, but carrying with him the curse of the Martyrs. The flight from the Kaiser's conscription and the long journeys across oceans to find space to farm and pray the way they wanted to, had settled in him as a need, once a year, to see new places, without people.

He had built his own canoe of wood and canvas. And he went north every August to follow the old trade routes toward Hudson Bay, the rivers charted by men who disdained farming, who worshipped—the few times they did worship—in dark, incense-filled cathedrals.

Tired, close to the end of his last trip, he misread a map and followed the river past a bend, running what he thought a small rapids. A waterfall, just a few feet high and fairly quiet in this time of low water, swept him down and broke his canoe. His foot caught between two rocks, and the water pushed him under. His parents hope that in his last moment, in the water churned white over rocks, he saw something—austere, gentle and unnamed—waiting.

III
Life Science

So Far North

On my only visit to New Orleans, I spent most of five days in the West Jefferson Hospital. My best friend was a patient there, and he was dying. The doctor said he had a cancer that branched out from his chest cavity like a spider's web, threads running out and piercing his stomach, his intestines, wrapping around nerves and arteries. By the time I arrived, the medical staff had stopped trying to do anything but kill his pain.

Even though my friend and I seldom spoke about our emotions or our fears, we had worked out a plan: he would ask me to come down only when he thought he might be in the last stretch of his illness. He knew I didn't have much money and couldn't afford several trips. Our friendship was such that we both understood what this plan meant and how it worked, although we never actually discussed the details.

There had always been a reticence between us. His family had moved south from Canada to Indiana just a year or two before mine. The slight hint of foreignness had brought us together in school. Over the next twenty years we simply figured each other out, knew each other's stupidities and when to take each other seriously.

For most of the five days I spent in New Orleans, I sat with him. I took one afternoon off from hospital duty, borrowed my friend's car, and drove south out of New Orleans on Highway 23, down into the heart of Plaquemine's Parish. Highway 23 parallels the Mississippi River in its final run to the Gulf of Mexico. The highway starts wide and divided, almost a freeway, protected from the floods of the river by high dikes. Behind them the river is broad and slow, its water dark brown, almost red. Farther south, the river spreads into salt marshes, and the highway peters out.

Every time-and-temperature sign I passed gave a higher reading: 98, 100, 102, 103 . . . . The car had an air conditioner, but I didn't use it. I wanted to feel the heat, even though I didn't like it. Sweat dripped off my chin.

As soon as I was out of the city—driving south through fields of crops I didn't recognize and over small streams that had been dredged into straight lines to drain the wetlands—I began to see birds I knew primarily from field guides: mockingbirds, boat-tailed grackles, tri-colored herons, cattle egrets, black vultures.

Before the highway ran itself out in the bayou, when it was still quite wide and busy, I pulled over to check a small drainage ditch. I left the car on the side of the road. A bird flew up as soon as I got close to the water. A shore bird. Thin. Long bright pink legs streaming out behind it like exotic plumes. It flew down the ditch about fifty or sixty yards, landed on the edge, then waded back into the water. Its thin body—a striking pattern of black and white, a long neck, needle-like bill—looked comic stuck so far above the water on those long legs. A black-necked stilt. My first. I hadn't expected it. I watched it for half an hour as it reached down from its awkward height to pick through mud. I could almost ignore my sweat.

Finally the stilt disappeared behind a clump of rushes, and I walked back. Just as I arrived, a cop pulled over behind the car. I nodded to him and waited.

Even though he looked to be about my age, he climbed out of his squad car slowly, painfully. He seemed like a Hollywood version of the southern lawman: mirrored sunglasses that reflected everything back at the suspect; a spreading belly that stretched out his shirt; sweat stains under his arms, another running in a V down the line of his buttons; heavy black boots; a billy club dangling from his waist.

"Whatcha doin', son?"

"Nothing, sir. Just stopping to look at some birds."

"Whaa?"

"The birds, sir. I've never seen anything like them."

"You a Yankee, boy?" He might have smiled, but I wasn't sure. He looked as if he could be in a newsreel standing in front of a courthouse keeping black people from registering to vote or in a scene from a fifties movie, in the dark corner of a jail, beating a kid from one of the area's more successful families, a kid who had turned to endless drunkenness and maudlin self-recrimination.

"No, sir. I come from so far north we call everyone south of us Yankees."

He didn't look amused.

"Then where'd'ya git the car with Louisiana plates?"

"It's my friend's. He's sick. I'm just visiting." I thought I sounded calm, thought I might even be appealing to his sympathies.

"I wanna see some I.D., son. Four pieces of it. And I want pichurs on two of 'em."

I gave him my Michigan driver's license, my Alien Registration card, my Ann Arbor Public Library card and my Wilderness Society credit card. He went back to his car. I stood sweating in the sun. The black-necked stilt flew up from the drainage ditch and crossed the road. Even though the bird flew right over me, I didn't lift my binoculars. I didn't want to look suspicious. Still, I noticed the bright pink legs streaming out behind.

"Guess yer OK, son," the policeman said when he brought my identification back. "Least we don't have anything on ya. Yet. But move along. An' leave our goddam birds alone, ya hear?"

"Yes, sir," I said. "Yes, sir. And thank you."

Life Science

It seems impossible—hearing the loud notes of a Kirtland's warbler in early June and finding the male easily (its yellow breast brightened by the sun) as it perches on the top of a small jack pine—impossible to believe this may be the rarest surviving songbird in North America. Its rarity has filled both the scientific and popular ornithological literature with descriptions of calls, plumage, or of stragglers outside the usual range. Although fairly large for a warbler, it is a small bird clinging to life in the world with as few as two hundred breeding pairs.

The potential extinction of the Kirtland's may be due to its own inability or unwillingness to adapt to any but the most specific conditions. It nests only in the blueberry, aromatic wintergreen, bearberry, sheep laurel or sweet fern that grow beneath young jack pines. The stands of pine must be large, at least a few hundred acres, and the birds move on once the pines are twenty feet high and their branches cut off the light that feeds the understory. The birds nest over a porous soil classified as Grayling sand that is found only in thirteen counties in the northern part of the lower peninsula of Michigan.

A specimen was collected from a ship as early as 1841 near the Kirtland's winter range in the Bahamas. The first American specimen was shot near Cleveland, Ohio, during the migration on May 13, 1851. Although the nesting area was discovered by two ornithologists from the University of Michigan (E. H. Frothingham and T. G. Gale) on a fishing trip in Oscoda County in 1903, the nesting habits were not thoroughly described until the early 1920s.

Then a young man from Chicago visiting his parents' summer place near the Au Sable River, found a nest.

Ornithology, like poetry and mathematics, is a field where the young and self-educated, excited by early abilities, often surpass their elders. The young man spent days watching the Kirtland's nest from a distance. When he moved in close, the male landed on his thigh and pecked his leg. The adult birds were disturbed by his presence and stopped feeding their young. He killed mosquitoes and horseflies and fed the fledgling warblers himself. Early in 1924 he published his observations in the prestigious ornithological journal *The Auk* (Number 41, pages 44-58) under the title "The Kirtland's Warbler in its Summer Home." Imitating the lettered names of his older co-enthusiasts, he signed his article "N. F. Leopold, Jr."

A few weeks after the article appeared, pursuing a jumbled mix of scientific curiosity and haphazard readings of Nietzsche, Nathan F. Leopold and his friend Richard Loeb kidnapped a child, Robert Franks, and killed him. They wanted to observe death. They stuffed the body in a culvert by Wolf Lake Park in Chicago. When the police questioned Leopold about a pair of eyeglasses they found near the body, he said he had lost them chasing a Wilson's phalarope, a bird rarely seen in that region.

The Disappearance of the North American Wood Warbler

In June I realized that I hadn't seen any warblers. Not one. Not even a yellow-rumped. Or a common yellow-throat. Certainly nothing like a golden-winged or a worm-eating. Of course, I hadn't been out often this last spring, maybe four or five times, and, when I had gone, I did it more for the exercise, the relaxation, than for any serious watching. My lists have been poorly maintained recently.

Usually I can see at least a few warblers even in my back yard. Magnolias, for instance. A black-and-white. A chestnut-sided. Yellow-rumped, certainly. And one morning a few years ago, just before dawn, when I couldn't sleep and was wandering around the back yard in my bathrobe, a male redstart fluttered around the corner of the house and actually landed on me for one frozen moment of absolute warbler terror. I looked down at him perched on my belly as he flashed the orange-red patches on his tail before pushing off and flying up over the house.

This year I didn't see any warblers. Not any of the common ones, not a rarity, not an accidental. Nothing. When I finally realized it, in June, I blamed myself, my lack of attention.

After a few weeks, however, I began to suspect that there might have been another explanation. Perhaps some previously unknown limit had been passed and one too many trees had been cut, destroying forever the warblers' winter range. Perhaps some kind of horrible gas had been released in one of the wars, killing nothing but warblers.

The newspapers didn't mention it. I looked for articles headlined "Scientists Search for Missing Birds" or "Tiny Transients Terminated." But nothing. I went to the library and checked the recent issues of *American Birds*. Not a line.

I took two days off work to look in the hedgerows where I knew yellows nested. Nothing. I went to marshes where common yellow-throats or black-throated greens had been in past years. Nothing. No warblers, anywhere. And no one seemed to realize it. I told my wife. She said I just wasn't concentrating. I mentioned it at work. No one was quite sure what a warbler looked like. I told them that this might be it, the first indication of the end. If the warblers could disappear, so could we. So could anything. They didn't believe me. They thought it would have been in the papers.

So I kept waiting. And looking. Every chance I could get. Even into the fall, when the migrations should have brought a palm or a bay-breasted back through. They would have been duller then, but still obviously warblers. But I didn't see any. Not one. And now it's November.

Lament for the Crested Shelduck
(Tadorna cristata)

They're gone now. The last record of them is from North Korea, March, 1971. Two males and four females were swimming in the sea at the mouth of the River Pouchon. Unmistakable. The males: glossy green, almost black on their heads, along the crests that were like manes on their necks; glossy green on their breasts; necks and cheeks paler, gray. The females: white necks; distinctive black and white masks. The bills and legs of both, a brilliant red.

They bred on quiet patches of rivers in the moist forests of northeast China, North Korea, and a small section of Siberia. They probably wintered along the coasts. Only three specimen skins exist, all in Korean museums. No seasonal variation or juvenile plumage is noted in the books. Their call—if they had one, and they must have had one—was never recorded.

Chekhov may have heard them when he floated east down the Amur in the Spring of 1890 on his way to Sakhalin, the prison island. In his long letters to his sister, Masha, or to his editor, Suvorin, he described the thousands of ducks and geese, flying and calling and scurrying from the Russian side of the Amur to the Chinese. The crested shelduck may have been there, and Chekhov might have recognized it from an oriental screen he had seen at the Countess B.'s country estate south of Petersburg. He didn't mention it. The birds would have been part of the mist in the evening, swimming off from the riverboat, reclusive and wary of men.

Cottonmouth at Turkey Creek

"A cottonmouth," Georgia said. "Cottonmouths can kill."

Lee hid in the doorway. Christine joined him. I tried to be brave and help Georgia identify the snake's field marks.

"A cottonmouth," she said again.

It spiralled up one of the columns supporting the back porch of the old house above Turkey Creek.

Georgia and Lee were checking on the place, periodically, for a friend in Charleston. The house was rumored to be haunted. A childless couple had lived in it for over half a century, and apparently nothing, not even death, could move them. The woman died while out picking strawberries in her hidden patch a few hundred yards upstream. Her husband carried the body back in a wheelbarrow, then had a stroke a week later.

We hadn't seen any ghosts.

We'd come at sunset to sit on the old bridge over the creek. We drank whiskey and watched the stars come out. We didn't check the house until long past nightfall. When Lee got the back door open and switched on the porch light, we saw the snake. We had passed right under it.

"If there's one, there's two," Georgia said. "Cottonmouths smell like cucumbers. Does anyone smell cucumbers?"

The snake was headed for a barn swallow nest under the eaves.

"If there are babies in there, it'll get them," Georgia said. "Should we do something?"

No one volunteered.

The snake moved slowly toward the nest, one foot, two, then three.

We talked about finding a broom and knocking it off, saving whatever was in the nest. But we were afraid the snake might spiral back on the broom, biting our hands, or, worse, landing on our heads. We cracked nervous jokes about first aid.

"Oh, well," Georgia said, "it's the way of the world, I guess."

We all agreed.

The snake arrived at the little mud cup nest, stuck his head in and pulled out one baby barn swallow. He started gulping it down, head first. Featherless wings flapped at the corners of the snake's jaw, then disappeared. The snake curled above the nest, looking at us, his red tongue darting. He went back into the nest three more times and pulled out three more young swallows. We watched the bulges ripple down his body.

When he finished, he extended slowly across the ceiling of the porch, at least five feet long and clinging, somehow, to a crack between the slats. When he finally fell, we all jumped into the house, knocking each other over and into the kitchen.

Back at Lee and Georgia's, we identified the snake in their field guide. Kingsnake *(Lampropeltis getulus).* Big and fierce looking, but nothing like a cottonmouth. Harmless.

The War on the Bats

Russ was afraid of heights. Lizzie was eight months pregnant, and they were trying to fix up their new old farmhouse before the baby was born. I wasn't working at the time, so I volunteered to paint the peaks.

I'm confident of my ability on ladders. Twenty-five or thirty feet off the ground, where most people start feeling their calves shake, I come into my own, leaning out so far people watching below gasp.

I spent the first day scraping and wirebrushing the old paint, knocking off wasp nests and repairing a few loose pieces of molding. That evening, after I'd showered away the worst of the grime, Lizzie took me on a tour of her garden. She had planted it in what was once the barnyard, although the barn had burned down a couple of generations earlier. The garden was doing well: corn, potatoes, lettuce, squash, pumpkins. Swallows swooped around our shoulders as the sun set over the cornfield behind us.

When we turned back to the house, I looked up at the peaks I would prime the next day. Spiralling above the roof of the farmhouse was a dark cloud of bats. Hundreds of them.

"My god," Lizzie said. "I've never noticed them."

We called Russ out from the kitchen where he was cooking a venison roast.

"What are we going to do? They must live in the attic. We can't leave them there. They carry rabies."

We discussed our options.

Russ and Lizzie could wait a few months, hoping that these were the kind of bats that migrated. I could come back in November, try to find whatever hole they were using and seal it. But Russ and Lizzie's baby would be born by then, and they didn't like the thought of hundreds of bats climbing around in the attic above their infant. Or I could find the hole tomorrow, seal it and hope most of the bats would find some other way out. If we made life difficult for them, perhaps the bats would swarm off to some other roost, a hollow tree or an abandoned barn. We decided that this would be the quickest solution.

The next morning I extended the ladder a bit farther so I could climb off it and straddle the peak of the roof. The bat hole was just a few feet back from the edge. Old shingles had fallen away and the beam below had rotted. The bats had easy access to the attic. I put in a few scrap pieces of two-by-four, making them as snug as possible against the remaining pieces of good wood. After packing the hole with fiberglass insulation, I covered the mess with shingles and roofing tar. I stopped often and looked out over the cornfields and pastures to the woodlots. A few white-tailed deer disappeared into the trees; a turkey vulture turned slowly high above the corn. I climbed off the roof and painted into the evening.

By the time I finished the undercoat, Russ and Lizzie were beginning to worry about the bats.

Inside the house we could hear them squeaking in the attic. They sounded frantic. I went outside to see if any of them were escaping. My patch job on the roof was holding. No bats. Around back, however, was another story. The bats had moved through the attic, down between the walls, and were squeezing out of a small hole above a first floor window. One at a time. Then flying off into the evening.

I checked back at the hole every few minutes for the next couple of hours until it seemed as if all the bats were out. I stuffed as much insulation as I could into the hole and went back inside to relax.

As soon as Russ and Lizzie and I sat down to play a round of three-sided Chinese checkers, a few bats still stuck in the attic started descending through the overhead light fixtures. I ducked. Lizzie ran for the bathroom. Russ grabbed a tennis racquet from the closet and I picked up a broom. We started swinging as the bats circled the living room. The first one Russ hit was flung across the room and landed on my shirt. I felt its claws scrape my belly. I looked down into its tiny teeth. It squealed. I think I screamed. I brushed the bat off and ran to the bathroom to hide out with Lizzie. We didn't come out until Russ assured us that he'd killed them all and had thrown them into the back cornfield.

The next morning a solitary bat was hanging upside down under the north peak of the house fifteen feet away from the upstairs window. The day was already hot, and I had planned to move around the house in the shade. I painted the west peak first, but when I was done the bat was still under the north side. If I disturbed it, I thought it might fly off, leaving the peak free to paint. I took Russ's softball from the closet. Outside, as I aimed for the bat, it occured to me that I might miss and throw the ball through the window. But, no, I told myself, even I didn't have that bad an aim. I wound up and threw the ball through the window. I spent the rest of the day repairing it. The bat didn't budge. The next morning it was gone.

Two nights later, after the peaks were painted and just before I planned to leave and head back to the city, we heard a scraping in the wall. Russ grabbed his tennis racquet, but no bats appeared. Then we heard a faint flapping noise from the living room.

Down in a far corner, reaching up between the wall and the floorboard molding, was a bat wing, very small, pink and hairless. It fluttered faintly, repeatedly, against the farmhouse wall.

The Hope of the World

I've never studied butterflies. Like most people I can recognize a swallowtail in the garden. I know that the white cabbage moths—the ones that always seem to be in pairs fluttering across meadows in August—are moths, not butterflies, although to explain the difference I would have to check my wife's field guide.

Last summer a friend told me that most of the small orange and brown butterflies we see around here are called fritillaries. I don't know where that word comes from. I should look it up. Fritillaries look like little monarch butterflies, and everyone knows about monarchs.

Monarchs migrate. In late September I can stand at the window of the shop, nestled away on the second story, and see monarchs flying south down State Street, fifteen feet above the pedestrians.

Once, in early autumn, I was on a small peninsula that sticks out into Lake Erie and found thousands of monarchs clustered on the last maple at sunset. They hung together in orange clumps, so many that their weight bent the thickest limbs.

A couple of weeks ago, just after a September storm, my wife and I spent a weekend along the northern coast of Lake Michigan. We found several monarchs that had been blown into the water during their migration. Some were dead. Others had clung to bits of seaweed or twigs and had drifted onto the beach. My wife picked up these butterflies, carried them above the shingle, and set them gently on the dry sand.

IV
The Blood Trail

Bear Stories

1.

Stanley Stewart's aunt read about it in the Calgary paper. A tourist from the States wanted to take his daughter's picture with the black bears at Banff. He coated her hands with honey. He stepped back, the child reached toward the bears like her father told her to do, and he snapped a picture of a bear eating his daughter's hands.

2.

Never run from a grizzly, old John (who claimed descent on his white side from headless Queen Anne) told me. They'll beat you every time. You can try to climb a tree, but any tree you can climb, a grizzly can too. He'll just walk up it like a ladder. No. The best thing to do is just stand there, and try to get up a big gob of spit in your mouth. Just before a grizzly gets you, he'll stand on his back legs and roar. This is your only chance. When he roars, take that gob of spit and spit it right into his mouth. If you get it back against his throat, he'll turn around and take off. It works better if you're chewing tobacco.

3.

LaVerne told how his first grizzly charged after the first shot. Since a grizzly's eyes are bad, LaVerne figured she was going for the smell of gunpowder. She came a hundred yards straight for him, LaVerne shooting the whole time. The bear dropped ten feet from his blind. When he gutted her, he found five of his bullets in her heart.

4.

John and I came back to the main camp after a few days cutting horse trail and found our cook tent ripped to ribbons, the table torn apart, even the stove knocked on its side. A few minutes later I saw the outline of a black bear sitting in the brush at the edge of our clearing. He must have been accustomed to people, maybe to dumps at mining camps or to other hunting outfits. John gave me the rifle, a small bore goat gun but the only one we had at the time.

"Your first bear, " he said.

I tried to aim behind the front leg. I squeezed off the shot slowly, like I'd been told. And the bear howled like a baby, or maybe more like a rabbit that has just been snatched by an owl. He took off through the woods, breaking branches and tearing up bushes. We waited a few hours, hoping the wound would stiffen him up, then followed the blood spots through the forest. About half a mile from camp we found where he had holed up, the ferns pressed down and covered in blood, but the bear was gone. The blood trail gave out soon after.

5.

It was a Sunday. I took my lunch and binoculars, but no gun. I followed a stream up the mountain behind camp until I reached the spring that fed it. Then I crossed a couple of ridges and moved up through the last scrub firs at the tree line. I kept climbing. The only things growing up there were small mountain blueberries. They were ripe and close to the ground. I knelt down and ate them, picking each berry carefully and wiping off the dust of the mountain. I continued climbing until I reached the last high ridge. I sat there, ate my lunch and looked out over the hills and forest below me. I knew our camp was somewhere down there on the edge of the lake, but I couldn't see it. The lake went off to the east for many miles. I could just make out the small valley where it emptied and where the Yukon River started. I pulled my hat over my eyes, lay back in the blueberries and slept.

As soon as I woke up, I sensed I was not alone and might be in trouble. I raised my hat, then my head, slowly. Below me, back down the mountain maybe a hundred and fifty yards but well up from the tree line, a lone grizzly was working his way across the mountain. The fur on his hump caught the sun and seemed almost golden. He looked larger than any animal I had ever seen, bigger than a moose, much stronger than the workhorses on my grandfather's farm. He was raking his claws through the blueberries, pulling out large clumps of them—roots, leaves, dirt—and stuffing everything into his mouth. I could see his teeth even without binoculars. The breeze was blowing up the mountain from the lake, but I still tried not to breathe. The bear didn't turn, didn't look up, just berried his way below me for half an hour. After he had disappeared around a far ridge, I lay, half afraid and without moving, at least an hour longer.

On the Way to Moon

LaVerne decided that the smaller camp up at Moon Lake needed an extra person. Delos and Jess were busy with the hunt. They could use someone to help cook and chop wood. And they needed someone to clear the upper stretch of the new trail.

Delos had always taken the horses up to Moon and then brought them back out over the high passes, where he spent two or three days above tree line. In June slushy patches of old snow made the horses slip or stumble. In October, when he brought them down, Delos sometimes got hit with severe storms, even blizzards. He needed a trail that would be quicker and safer.

Earlier in the summer John and I had spent a week trying to find a passable trail up to Moon through the valleys. We'd succeeded, but the trail would need a lot of work, clearing snags and blowdowns. LaVerne sent me up because I knew the trail and would be more useful at the smaller, higher camp. That's what he said, anyway. But he may have just wanted me out of his hair, some place where he didn't have to worry about me saying or doing something stupid.

By the time we had closed up camp on the big lake and boated down to the beginning of the new trail up to Moon, it was past four. Summer nights that far north never turn completely dark. I wasn't too afraid of losing my way in the night or the forest.

I did get a bit nervous when LaVerne insisted I take a rifle—"just in case." He gave me an old gun I had never seen anyone shoot, spent a couple of minutes oiling it, and used a small stone from the lakeshore to knock the sights back into line. He told me to shoot the gun only if necessary, gave me five shells and explained how to release the old-fashioned safety.

The first part of the hike was easy. I had walked it just a few weeks before, and we had blazed the pines and firs every twenty paces. The path climbed steadily and had several steep spots that I thought might give the horses some trouble. In a few places dead trees had blown across the trail. I would have to spend a couple of days here with a chain saw before the ponies could make it easily with their packs. Whenever I climbed over anything rough, the old gun smacked against my back.

The forest was quiet. There were signs of deer, moose, wolf and black bear, but I didn't see any or hear anything running off through the undergrowth. A few smaller birds sang, and in one clearing I looked back over the big lake and saw a golden eagle soaring below me. Of course, I couldn't see any sign of LaVerne and the others in the yellow outboard.

After three hours I stopped to eat the lunch I had packed. I studied the rough map I had been given and figured I was about half way there. Soon I should reach the end of the blazed trail and be in the high valley where Moon Creek ran down from the lake. The valley was crisscrossed with animal trails. All I would have to do was find a good one, follow it until I had to ford the creek, climb a low ridge on the opposite end and descend to the little cabin that would be easy to see, sitting right at the edge of the lake, the only man-made structure anywhere within thirty miles.

By the time I had finished eating, the shadows of the northern half light were filling the trail. They were unlike those cast by the sun. They didn't stretch, but just clumped down in dark blobs on the landscape. Although I could still see clearly, the trail, the pines, even the mountain peaks around me were all covered by the misty evening light of the north. The two men eight or nine miles ahead in the cabin at Moon were probably already in their sleeping bags. The men behind were far down the big lake by now, maybe at the road forty miles away.

When I got to the edge of the treeless valley, I was afraid. It looked endless in this light, stretching off into mist and the half dark. Just below me three moose grazed on the scrub bushes that grew by the edge of the creek. I stepped five paces into the valley and looked down at the tracks in the mud. A fresh grizzly print—the slash marks of its claws digging into the ground around the paw—filled the center of the trail. I bent down and stretched my hand out in the print. The circle of it was far outside the reach of my fingers. I swung my rifle around, fiddled with the safety until I was fairly sure I had released it, and marched down to the creek.

I didn't stop to look for a ford or to take off my boots and socks. I just walked in. Knee deep. Then up to my waist and freezing. Then chest deep, and I held the gun above my head. The current pushed at me and my feet caught in the rocks. I didn't stop. When I climbed out on the opposite bank, I pointed the gun ahead and set off on a sloshing trot toward the cabin at Moon.

Horse Stories

The hobbles were about a yard long and made of soft rope. Small loops at each end fastened around the back ankles. Once they were hobbled, the horses could still move, although not very fast or very far. Delos told me they could rear if they needed to defend themselves from wolves or bears, and they could graze through the night without much discomfort. The hobbles wore into the skin on the horses' legs. We had to grease the sores each morning to keep out infection. We left rope halters on the horses so we had something to grab when we got close enough. In the morning we carried out oats and gave each horse a handful. They came to expect the treat and were usually easy to catch.

On the first morning of hunting season, Delos took me out to help bring back six horses. We left the cabin at 4:00 a.m. The sun, which had barely set that far north, was already turning everything gold. We found the horses about a mile away. Although they were stocky bush ponies with thick and matted manes, all a dull gray-brown with only slight variations in their markings, they glistened like thoroughbreds at first light, the dew on their backs shining in the sun. They gathered around us when Delos took the oats from his pack, and we picked out six. After we freed the rope leads from the halters and took off the hobbles, we started back.

Delos was in a hurry. He wanted his breakfast. He assured me that the horses were easily ridden bareback and that they were calm enough to allow the rope from their halters to work as reins. He would ride one, herd two and lead a fourth. I would ride one and lead one. He set off first, letting his horse pick through the scrub until we emerged on a game trail that led back toward the cabin.

I had no trouble following him on the first part of the ride. But when he had a clear stretch, Delos allowed his horse to break into a trot. The gelding I rode followed. The uneven ride soon had me bouncing all over his back. I squeezed my legs against his sides, but it didn't help. I wanted to grab his mane, but one hand held the rope that served as a rein and the other held the halter of the horse I was leading. Whenever the second horse slowed, I had to yank him along. At one point he tried to bolt, almost pulling me off. Sideways, I clung to the neck of my horse, my left leg barely draped across his back. But I didn't let go of either horse, and I didn't think Delos saw me slipping. Soon he slowed, and I was able to scramble back.

The trail crossed a small stream, not more than three or four feet wide, just a couple of hundred yards behind the cabin. Delos's horse came to the stream and jumped it without breaking stride. Mine came to the edge and stopped. He looked down at the water and tossed his neck around. I kicked him in the flanks, cajoled him, yelled at him. Just when I thought I would have to climb down and pull him across, he jumped. The horse I was leading didn't move. I did not let go of either rope. For one clear moment I was suspended in midair. Then I was sitting in the middle of the stream, a rope in each hand and a horse on each side looking down at me.

A few days later Delos asked Jess and me to ride out of the small valley that held Moon Lake and scout for game to the east. I found two horses and brought them back to the cabin without incident. We saddled them and set off at a leisurely pace on the trails that led around the base of the mountain behind the cabin and up into a far valley. Several moose grazed along the stream banks. Jess pointed out the wolf and bear scat on the trail. Whenever we stopped to scan the surrounding peaks with our binoculars, we saw mountain sheep and goats.

On the far side of the valley we rode as high up a small mountain as the horses could go, then took off the bridles and used the halters to tie the horses to a bush. We climbed several hundred feet higher until we found a comfortable spot on the alpine tundra. We spent several hours scanning the valley and the mountains across it, keeping a rough tally of the animals we saw and noting their locations. Late in the afternoon, we climbed back down to the horses.

After we put the bridles back on, Jess mounted, took the end of the rope halter—knotted to keep from fraying—and gave his horse one flick on the rear. The horse took off on a smooth gallop down the mountain trail.

I tried the same thing but missed. The knot went under the horse's tail and lodged high up between his legs. He began rearing and bellowing. I managed to stay on, even while I reached behind trying to dislodge the knot. At one point, as the horse continued bucking down the side of the mountain while I held on to the saddle horn with one hand and reached back tugging on the rope with the other, I looked up. Jess sat on his horse a few feet off the trail, smiling as I bucked past.

When Delos and Jess took the hunters out on long trips, I was left alone for several days. My responsibilites were light, and I was able to read several of the western novels that had been left by previous clients. Horses, guns, bad guys, lawmen, and the endless empty land just there for the taking. Between novels, I chopped wood for the cookstove, cleared a few trails, then went out to check on the horses and grease their hobble sores. The grease was fat drained from our frying pan. I thought the aroma would carry for miles.

One night, soon after I returned to the cabin and was settling down with a Zane Grey novel, I heard the chorus of a wolf pack. Although the howls—plaintive, hardly menacing—echoed through the valleys, I was certain the wolves were close, possibly even at the place where I had last greased the horses.

I wanted to wait until Delos and Jess returned before checking, but the wolves could have destroyed the whole herd by then. They could have done it by now. So, as soon as the morning brightened past the heavy shadows, I loaded my rifle and walked out. At each turn of the trail I expected to find the half-eaten carcass of a horse, blood splattering the bushes, blank eyes staring up at me. But all the horses were grazing in the meadow where I had left them. When they noticed me, they hobbled forward for their oats.

Cutting Trail

Three days before the bush plane was coming to fly me out, I put an ax in my foot. It was silly, of course. Stupid. I'd set off from the cabin at Moon Lake early, even before Delos and Jess had left with the hunters, so I could work down the new trail, cleaning snags, chopping through blowdowns, and making sure the blaze marks were clear. I was hoping I could get through four or five miles that day.

I hadn't gone a mile before I did it. One of the scrubby alpine firs had blown over near the trail. I thought there was a chance one of the branches could snag a leg or a horse pack. I wanted the trail to be clean so Delos would have easy going when he brought the horses down in October.

I braced and swung. But I wasn't thinking. Instead of hitting the branch at the hard point where it grew from the trunk, I hit it a yard or so out. The ax glanced down the branch and went into the top of my left boot. It didn't hurt, and I assumed the boot had stopped the blow. I raised the ax to strike again.

Then I felt the wetness around my foot. Blood bubbled through a small hole that cut neatly across the laces and the leather tongue of my boot. I didn't look again. I started running back to the cabin.

About halfway there I felt the blood pounding in my temples. By then I could feel the pain, each heartbeat striking hard on the top of my foot. I slowed to a walk, then a hobble. When I reached the cabin, I saw that the horses were gone. Delos and Jess had already taken the hunters out into the mountains. I would be alone until far into the evening.

I sat on the big log I used for splitting wood and cut my boot and sock away. They were both wet and dripping with blood. I tried to press on the cut but the blood pushed up through my fingers. I took my belt off and pulled it tight around my lower calf. But still the blood flowed, dark red and thick.

The shore of Moon Lake was twenty yards away. We took our drinking water directly from it, and the water was so cold it often gave me a headache when I drank too quickly. I walked to the lake, then into it. Just a step, and already the water was knee deep. The cold clamped around my legs. The blood rose lazily into the water and drifted off.

I stood there until my legs were numb and the bleeding stopped. Through the clear water of Moon, I looked down at my foot, pale blue, with an inch-long purple wound in the middle, a small, open mouth.

The Persian Doctor's Second Daughter

Southern Indiana was almost the South, or closer than Marc and I wanted to be in late June. We were always hot. The bugs droned on all night in what seemed a tropical noise. Nighthawks punctuated the dark with their calls.

We were there to paint an apartment complex, a one hundred unit affair stuck in a cornfield. It was owned by a Persian doctor, a friend of a friend, and was a place that could have been anywhere, plain apartments strung together with a few odd angles to make them look interesting. In the middle was a poorly maintained swimming pool next to a clubhouse that the tenants could use, although few did. Marc and I slept there.

The doctor's second daughter and her husband were the managers of record, but we never saw them do much. A third of the apartments were empty. The coat of paint we were applying was meant to cover the most obvious problems. The husband checked on us occasionally. Sometimes he brought us tall glasses of gin and tonic in the afternoon. Then we would climb down from our ladders and spend a few hours in the pool.

The doctor's daughter spent most of her time in her air-conditioned apartment watching television. She was friendly, although Marc and I never found any subject that would hold her in conversation. She was good-looking, perhaps a bit aloof. She looked bored; in fact, she looked as if she could never be anything but bored. Neither of us found her attractive. When Marc and I talked about women we discussed our fiancées back north, the ones we wrote long letters to and called once a week from the pay phone in the clubhouse.

The summer moved on, and we moved slowly around

the buildings, slopping paint over mold and rotting wood, dodging wasps when we had to knock their nests from under the eaves. Evenings we stocked the clubhouse refrigerator with beer, then drank it standing on the second floor balcony looking out over the pool. We were usually there long after the last swimmers had gone back to their air-conditioning and late night television.

One night a couple of weeks before we finished the job, just as we were thinking about curling up on our bedrolls, that time of night when the bugs and birds were loudest but when the pool shone quietly below us, glowing in the green light cast by the one underwater bulb that still worked, the Persian doctor's second daughter came to swim. She smiled up at us, then dove into the water.

It's hard to explain what happened. Nothing happened. But whatever that nothing was Marc and I didn't talk about it until a decade later. The Persian doctor's daughter wore a modest black swimsuit. Her body stretched through the water. Her black hair flowed over her back. Her skin glimmered in the green light. At first she would stop at the end and pull herself partly out of the pool below us, let her head drift back, her eyes closed, before she turned and swam away. Her swimming barely disturbed the water.

I wasn't embarrassed. I didn't turn away. Marc watched too. This wasn't voyeurism. Or flirtation. The three of us were joined. Three bored people late at night in a hot climate surrounded by cheap and crumbling apartments. But one of us was bathed in green light.

Marc and I stood on the balcony. The woman swam below us in the pool, and the cicadas droned away, and the nighthawks screeched in the sky until dawn.

The Pool

He knew he wouldn't swim laps. He didn't want exercise. He wanted to relax. To feel clean. But since the firm was paying for the three months he spent in the apartment complex, 500 miles south of home, he would use the pool.

He started swimming during his first week. He spent each day going through the books of a small company that had hired his firm to do its audit and simplify its accounting. He worked long hours, finishing late in the evening. By the time he was back at the apartment, had called his wife and changed into shorts, night had fallen, and the outdoor pool was usually deserted. Only one of the underwater lights in the shallow end worked. The light was misty and filled with shadows in the deep water.

He had been a strong swimmer since childhood. He couldn't even remember learning to swim. It seemed as if swimming were something he could always do, and he did it much better than anyone expected. He had never been athletic, had never particularly enjoyed playing or even watching sports. But he could swim. The childhood plumpness that he had never lost, that had moved straight toward a middle-aged spread, made him surprisingly buoyant. But back home he had never spent much time in pools. There the small glacier-formed lakes, with pines growing right to the edges between the piers of cottages, the lakes that were cold and very deep just a few feet from shore, were the only places he felt comfortable swimming. He thought pools were for city people who couldn't get away. They were crowded and purified with chemicals that hurt his eyes and made his skin itch. When the firm said it would provide an apartment with access to a pool, he first thought he would never use it.

But that far to the south, the summer heat drained him. The apartment and the office he worked in were air-conditioned, and the air smelled stale. The drive between the two was enough to coat him in sweat. A shower didn't refresh him. He went to the pool the third night he was there and returned every night afterward. The chlorine bothered him, but he found that the coolness he felt later was worth the irritation. He tried to keep his eyes closed.

At first he had been satisfied with a gentle breast stroke or with floating on his back, occasionally squinting up through the haze at the stars. Then he began going underwater, pushing himself away from the cement edges, arching his body down toward the deep end, not kicking, simply gliding through the water as long as he could hold his breath. On the second night in the pool, he discovered a new way of swimming, one that felt immediately right, the perfect way to relax.

He thought this new way might look childish. But it was late, and the poor light in the pool would hide him from anyone just passing through the parking lot. Soon he was not even bothered by the occasional other swimmers, and they seemed to recognize his mood. They left him alone.

He would tread water in the deepest part of the pool, then put his feet together, bring his hands up from his sides in one smooth stroke, and move slowly toward the bottom. The water was just deep enough that he felt a slight pressure on his ears. When his feet touched the drain, he drew his knees up to his chest and began drifting upward, floating toward the air. He kept his eyes shut. Occasionally—and, at first, these were the best moments—he lost his sense of direction, felt weightless and helpless in a cocoon of water. When his head or shoulders broke the surface, he fluttered his hands a couple of times, took a deep breath, then pulled himself back toward the bottom.

After a few evenings, he became distracted from the pleasure of the movement by sounds in the water. He first noticed them on one of the nights when other swimmers were in the pool with him. While he was on his slow ascent, he heard their voices, muted and wordless. They seemed metallic. The splashing was amplified. It carried through the water like the recognition of a dull pain. When they left, he heard small waves break against the edge of the pool.

The next night he listened again, even though there were no other swimmers. He heard the sounds of the pool, the hum of its pumps, the occasional chime of what might have been a small bell, the swish of bubbles that rose when he exhaled. He began to hear a music in the noises, each sound playing with or against all the other sounds. He thought the music might have a pattern, a repetition that he couldn't quite understand.

As early as the third week he began hearing other sounds, quieter sounds that came from somewhere far away. At first he thought he was just tired, that the sounds came back as a high-pitched memory of numbers he had counted during the day. But one night he began hearing them as a pleasant chatter. Like dolphins, or porpoises. The sounds became clearer as he listened harder. And he heard the calls of other sea mammals, the barking of seals and otters. By the end of his first month he could hear, faintly, even the keening songs of the great whales calling to each other through thousands of miles of salt water. The sounds in the pool became clearer each night. He held his breath longer and tried to slow his ascent. The mix of voices from the watery world came up to him through the chlorine of the pool, past the pumps and drains, up the streams and rivers from the wide and distant oceans.